Cryptocurrency Mining 2022

The Best Guide to Mining Cryptocurrencies: Bitcoin and Altcoin Mining

Table of Contents

Preface

If you are interested in Bitcoin mining, I may be inclined to think that you understand what Bitcoin is. And while that may be the norm for most of you, I am sure there are some of you that are looking to understand Bitcoin mining as well as understand what Bitcoin is.

That way, those of you who have no idea how Bitcoin is used, can still get involved with mining it and be able to make it an investment that literally earns on autopilot. You see, once you set up the equipment and install the software, the mining is fairly straightforward as long as you understand your moves.

Bitcoin mining, Altcoin mining, Crypto mining, and Monero mining are all terms you will find in this book and they are not interchangeable. Bitcoin Mining will reference the particular nature of the process of Bitcoin, while Altcoin will reference the entire cryptocurrency universe minus Bitcoin. Cryptocurrencies or cryptos will refer to all cryptocurrencies. Finally, a reference to individual altcoins, like Monero or Ethereum — or whichever — will refer to the exact crypto that is being discussed.

The crypto mining process is not labor-intensive, but it is hardware-intensive and it depends on the amount of processing power that you can tie together to get the next set of hashes.

In a nutshell, mining is not so much about digging the earth and separating a mineral from the dirt, it's about solving complex mathematical puzzles and getting the solution to them before someone else does. There are two steps to this. First you have to get the function's solution and then you have to put it into the blockchain. The miner who is first to do it gets the reward.

The reward is in the form of coins. For every block that you solve, you receive a number of coins in return. The key to succeeding at this is to understand how it's done and the probability of success. Once you understand your chances, then you can understand how much to invest in the hardware.

The good thing is that once you get the hardware, you can mine for various cryptos and not just for one. We will show you how to set yourself up to be able to mine coins from diverse platforms.

Introduction

The industry seems to have settled in on two names to define the cryptocurrency universe. Regardless of the fact that there are more than 1200 individual coins in the market today with varying and diverse uses and distribution, we can synthesize it all down to two groups. On one side you have "Bitcoin", which stands there all alone. On the other side is what we call "altcoin" which is the catch-all term to represent all other coins that are not Bitcoin.

But we seem to get caught up with the currency because that is the visible front of the ecosystem. We see Bitcoin as money, or altcoins as money, and we forget that there is an underlying mechanism that goes along with each.

In this book, and in most academic and advanced discussions of cryptocurrencies and coins, there is a standard nomenclature that is followed. We will highlight it for you so that we are consistent throughout the discussion.

Cryptocurrency simply refers to the use of cryptography to create a unique and identifiable packet that can be used as a store of value or as a unit to exchange and barter. We call this currency.

It intrinsically does not have a value, other than the one that the majority of the universe agrees on. It's just like paper money that is printed and only has the value we say it does – well, maybe a tiny bit more complicated!

As for cryptocurrencies — or cryptos — there are a few parts to it that we will get into in Chapter 1. The first is a ledger, and that ledger in some cryptos is public, while in other, newer ones, it is not as transparent as the ledger Bitcoin uses. You'll also need to learn about a private key, a public key, and the concept of a hash. We will get to them in the following chapter.

But then after all that, we have what is called mining — and it is completely not what most people think! it is certainly not the process of taking a pick axe and striking stone to extract some secret code or valuable metal. Instead, it is about contributing your computing power to assist in the ongoing operation of the infrastructure.

Let me explain.

If you think about a currency, at the very basis of its existence its function is to be a vessel of value that can be transferred without difficulty and friction. You have a dollar in your pocket right now and you want to purchase a widget from the guy in front of you. When you hand over the dollar, the guy now has possession of the dollar - represented by that US Treasury's piece of paper. The paper itself has no value, it is no different than a piece of paper that you put in the printer.

But what gives it legitimacy and value is the fact that it is backed by the word of the sovereign nation. It is a contract between the nation and the holder of the note. But the paper itself does not have any value. If you shredded that paper it would not yield any value, unlike the melting of gold which will still give you value. That's why in the olden days, gold was used as the store of value. You could melt the coin and remove all markings of it, and it would still be worth a certain value.

This is called the intrinsic value. What is printed on the coin is called face value - since the value it is supposed to carry is printed on the face of the coin. If you then advance that a little further, think of a check that you can write whatever amount you like on, as long as it does not exceed the amount in your account. When the bank receives that check, it pays to the bearer the amount indicated on the check. If the person was not withdrawing cash, the bank just moves the amount from your account to the issuer's. That change is reflected in your account as a reduction, and indicated in the issuer's account as an addition. Yours goes down, theirs goes up.

That change is reflected in the ledger. In the early days of commerce and banking, ledgers used to be gigantic bindings of parchment, meticulously documenting the flow of money from one person to the next.

Let's say there were only three customers of a new bank and each had one hundred dollars in their account. The bank would keep the cash in the safe, and write in their big ledger. Mr. A has $100, Ms. B has $100, and Mrs. C has $100. If Mr. A gave Ms. B a check for $20 and she deposited that at their common bank, the bank would not bother going to its safe to take the cash out, they would just adjust Mr. A's balance by $20. So now he would only have $80, while Ms. B would be adjusted to show $120. The same $300 in total will still be sitting in the vault. The only difference is that the amounts of how much Mr. A and Ms. B could stake a claim to has changed.

What would happen if the bank suddenly lost that ledger? The money would still be in that safe — all $300 of it. But now the bank would not know who could legitimately stake a claim to that cash. That's the reason that ledger is important. It is the sole determinant of asset ownership. Once that ledger is lost, so is the ownership of the funds in the vault.

Not too long ago, the banks issued savings passbooks. So in case anything happened to the ledger at the bank, there was a duplicate copy of the transaction and the ownership of assets. You held that passbook and kept it in your safe. When you made a deposit, a notation was made along with the current balance, and the same was done when you made a withdrawal.

This was a little safer because now there were two records mirroring each other and in the event one was destroyed, the other could still be used to determine ownership.

So where is this all going with cryptocurrency? We are getting there.

If you think about it, the ledger is made from paper which in and of itself has no value. Its value is intangible because you could not take that ledger and sell it for the paper that is in it, nor could you sell it as a ledger. But as a keeper of records, it is invaluable.

On the other side of that equation is your savings passbook. If someone stole your passbook, and tried to empty the contents of your account, they would not be able to produce the valid picture ID that would be required to withdraw the cash. That picture ID is yours and can't be stolen because it has something unique to you — an image of your face — that can be matched to the picture ID. Together they unlock your access to your funds kept at the bank.

In time, passbooks were done away with and all you needed was your ID or an ATM card that came with a secure passcode. Combining the card and the ID, you could get your money.

So the elements in the entire process are the ledger, the passbook, the account number and the picture ID. These are all you need to create a secure chain of possession and transfer for an asset — in this case, cash. But it can also be anything. It could be the title of your car, deed to your property, certificate for your diamonds, the policy to your annuity or whatever.

Let's take this one step further. If you have cash in your bank, and that is the asset that we are talking about in the illustration thus far, then think about what was said earlier as to the value of the asset.

Let's take the US dollar for instance. Where does the US dollar get its value from? Well, it gets its consensus value from an intricate market-driven consensus mechanism. That is the Forex market. The US dollar gets part of its value from the consensus of the rest of the world. If the consensus is that the dollar is weak, then the market sells the dollar and the value of the dollar starts to depreciate with respect to a number of currencies worldwide.

At present, the value of the dollar against the British pound is approximately $1.39. That means for every British Pound that you wish to buy, you would have to pay $1.39. Or if you wanted to buy a dollar, you would have to spend 0.72% of a pound. The value of the dollar is market driven and market determined. No single entity or person can control that, theoretically.

Whatever we have that is of value is not represented by paper (called fiat) currency. The dollar bill merely represents it. Just like the photo of your dog is not your dog, but merely a visual representation. The wedding band you have on your ring finger, is not actually your marriage, but a tangible symbol of the intangible marriage. Value is intangible — no matter what you think of it. You can't really hold it in your hands. But the things that we use to represent it need to have some tangibility, some way for us to hold it or see it. That's why there was paper money and gold coins.

At the beginning of commerce, which is as old as civilization, ways to transfer value were created based on the need and the technological ability. When they first started commerce, trade was facilitated by barter. You wanted to buy my cow, and I wanted your goat, so we struck a deal and set the consideration as one cow in exchange for three goats. The value of our exchange was contained in what was exchanged.

But that presented a unique problem because, what if I wanted your goats but you had no use for my cow? So eventually a standardized system of exchange was created and each individual item could be purchased using one common item — currency. I could pay you currency for your goats instead of begging you to take my cow. You could then use that same currency to buy whatever you wished — butter, milk, or even jewelry for your wife. Everyone was happy.

The concept of currency worked as the medium to carry value. It was efficient, as long as I could see you face-to-face and give you the cash in return for the goods. And in those days, we bought things face-to-face.

Today, that has changed. We spend more online when purchasing goods, and that is only going to increase over time. When we buy something online, we may be sitting in Corvallis, Oregon, but the seller may be in Sydney, Australia. I want the item, and they want my payment but it takes a really long time to put paper currency in an envelope and dispatch it via the mailman. So then civilization came up with payment transfers and even value transfers like PayPal.

The difference between paying someone in person and paying someone online, was that one was face-to-face, the other was ostensibly long distance. One was instant, the other took days. One was free, the other cost fees. But more importantly, the process has been inefficient and this inefficiency increases costs of transactions.

We are at the same place in monetary evolution as we were when we transitioned from barter to currency. Except this time we still have currency, but the store of value (which we call currency), does not need to be tangible since it does not need to have intrinsic value anyway.

That brings us to cryptocurrencies, finally.

As currencies have evolved in the past based on the nature of the transactions and the technology that has grown around them, we are now at the juxtaposition of advanced technology and a new paradigm in consumer behavior. All signs point to the need for a new store of value. This time the store of value should be something that is able to lend itself to the kinds of transactions this generation is getting into — online transactions, while it is still able to conduct in-person transactions.

All these issues are adequately and elegantly addressed in the use of cryptocurrencies, and as such, the market for cryptocurrencies has exploded in the last decade. Yes, it is still new as an industry, but nothing can stop an idea whose time has come.

One of the elegant factors of cryptocurrencies is that it is decentralized. That just means that the currency is not regulated or controlled by any single actor. It is just as efficiently driven by markets. But that's not for the price alone; even the administration of it is driven by the participants.

For the rest of the book where we focus on the mining of Bitcoin, it is important to understand two sides of the cryptocurrency coin (please excuse the pun). The first side of the proverbial coin is that it is transaction-driven. The second side of the coin is that each transaction is the value of the coin. The coin is merely the vessel.

And with that, we shall move on to the technical illustration of the coin and the mining that underlies it.

Chapter 1 What is Cryptocurrency (from a mining perspective)?

In the crypto universe, there are two kinds of endpoints that are connected via a P2P protocol. One is a regular public node and the other is a miner's node. Think of it as a web, and every node as a junction. All nodes connect to a small number of other nodes directly, but eventually are connected to all nodes indirectly.

Nodes

A node is one instance of an address appearing on one computer. It is possible to configure one computer to have multiple nodes but that would not usually be something that the causal user would do. In this book, one node is simply assumed to be one computer at the end of a P2P network. A minimum of two nodes are needed to be defined as a network, As of this writing, there are approximately 12000 nodes on the Bitcoin network that are alive at any one point. There could be more, but many of them could be kept offline and thus not identifiable for the scan that was initiated to locate them.

There are two kinds of public nodes. One is a regular users node and the other is a light node. The former has full blockchain information. The latter has only the abridged version of the ledger, relevant to the coins that are stored in that account.

A regular node is then one of two kinds — one that has the full ledger, and the light one which only has the ledger of the coins that relate to it.

P2P

P2P protocol exists without the need for the internet but works better with it. A P2P network is just a network of computers that connect directly to each other using only one open port. There is no server needed for this relationship and it goes directly between the user nodes. That port on the node only allows the other person to send you specific data which is in the ledger.

So if you are worried about getting hacked or someone coming in and stealing information, don't be. That is still fairly difficult to do and doesn't happen unless the penetrator is highly motivated.

Your software compares the ledger that you have in your computer to the one that is being shared with it. If there is a difference, it then waits to see the ledger of a couple of other nodes.

Different cryptos handle the overall behavior a little differently from one another. But in general, you can set the number of nodes you connect to get updates in the ledger.

A Block

The information in the nodes is presented as blocks. A number of transactions go into a block. That number can fluctuate between 1000 and 2500. There is a website that tracks the average transactions in a block. It can even be as little as one — there is no limit to how many transactions can be contained in a block, but there is a limit to the size of the block. It is hard-coded that for Bitcoin, each block is no larger than 1 - 2 MB. Other cryptos will have other limits.

Once a transaction is placed in a block and that block is hashed and placed into record, it forms a chain with the block that preceded it and will continue to be part of a chain for the blocks that are added after it.

One block typically has the information of all the transactions that are included in it. It contains the hash of the last block, and a few more pieces of information like the node that calculated it and so on. But the important thing in the block is that it contains the hash of the previous block.

All the transactions in the block are put into a computational algorithm, meaning they are processed repeatedly until they produce a fixed string of characters. This is a cryptographic hash that only works one way, and it is not random. This means that if you place the same information into the block and hash it, it will always result in the same string. But if you change even one bit, the entire string of hash will change significantly. You will see this later in the book.

If A paid B, C paid D, and F paid G, then all those transactions are collected and placed in one block, and that block is added to an existing record of transactions.

The block is then hashed (we will see how that happens in the next chapter), and this hash then is entered into the public record that everyone has access to.

Once it is hashed and accepted by the network, it forms a chain because, if you remember, in the block there is the hash of the previous block, and in the hash of the block that comes after this, there will be the hash of this block. So one block is tacked onto the block that precedes it and that information can be tracked back all the way to the first block that was ever created — the Genesis Block.

Gossip Protocol

The P2P network works on an automated gossip protocol and it is constantly getting in touch with its neighbors to find out if they know what it knows. If the neighbor has less info than it does, it will transmit the new info. If the neighbor has new information, it will receive updates from the neighbor. It is a pretty efficient system because on average, each node connects at random to six other nodes. If each node connects to six other nodes and passes its info to them, and each of those six nodes are again connected to six others — and pass that info along — it only takes 7 (actually, less) steps to propagate the most current blockchain to the entire network of 12,000 nodes.

The propagation of data in a gossip network across a P2P framework allows the most current information to be spread across all online nodes within a few minutes. For any transaction to be confirmed, it takes about ten minutes to half an hour. So even if A was sending money to B — half the world away — the transaction could be complete in moments and confirmed in less than an hour.

A new node that comes online would have to download the entire ledger onto the computer, and this could take anywhere from 12 to 24 hours, depending on the traffic and the speed of your internet connection. Once you have the up-to-date ledger, updates are rapid and you can shut down your computer at any point. Then when you fire it back up, it will hunt down and seek the updates made since the last time.

The processes is automated for all crypto nodes.

Ledger

As you saw in the bank or fiat currency example with Mr. A, Ms. B and Mrs. C, there was a ledger that the bank held. Now you know that there is no central authority in cryptos, so who keeps the ledger? Well that's the part most of you do know! Everyone has a copy of the ledger and it is arranged as a series of transactions that are then linked to become a chain. Whenever a transaction is conducted, let's say, instead of Mr. A giving Ms. B a check for $20, he gave Ms. B a crypto coin for that same value. Then it is recorded in a ledger as a transaction and that ledger is shared by the entire network.

Unlike the example with the bank, this ledger is now distributed, rather than centralized. The immediate benefit of that is if any one party loses the ledger, there are numerous copies that will automatically be available to duplicate the information.

When your crypto app logs on to the network, it then checks on the status of the coins that you own and displays the balance on your screen. The thing here is that the number (the balance) shown to you is not taken from just what your node remembers it to be, it's taken from the entire, existing blockchain.

By doing this, every node is witness to your ownership of the coins that you possess. So what happens when you spend your coin? This is where you get to use your private key and the public key. But before we get to that, let me just explain what a hash is so that you can appreciate what a private key and public key are, and what mining is all about.

When you have a full understanding of a hash, you get the logic of what is going on when you start mining. Although, if you do not understand the whole thing with hashes or keys, don't worry about it. You will still be able to mine cryptos.

Hash

Get ready to be confused. There are a couple of different ways of defining hash. In conversational terms, hash just refers to the string of characters that is the result of smooshing together all the data that is being hashed.

But in actual fact, hash refers to the process where your data of any size is crunched up and returned to one fixed size of characters. This hash is typically on a one-way street, or what is known as asymmetric cryptography. That means you enter the source document and get a hash, or what is called a digest. But if you try to reverse that digest, you will not be able to find the content of the source document.

This is how passwords are stored on sites. When you set up an account, let's say your online bank account, you are asked to enter a password. That password is then converted to a hash and stored in the bank's encrypted drives. When you sign in to your account and provide the password, the password you provide is again converted to a hash and compared to the hash that is on file. If they match, you gain access. If they don't — well, you now the rest. The point is that what is stored in the drives is not the exact password so that in the event the bank's servers are penetrated by bad actors, the password file they get will just be hashes and nothing that they can reverse engineer.

That's the beauty of hashes. You can't take a hash and put it through an algorithm to get back the original text.

Point to Remember: Keep in mind that you can place however large a file in a hash generator, and when you run it, it will return a string of characters that look like this:

0306c2cdb84f97faa14618cabaf6c2bf

That is the MD5 hash for the entire lyrics to the national anthem. Go ahead and try it if you would like to see how it works for yourself. You can go to

Google and look for an "online MD5 hash generator," and when you find one, place whatever you want in it. Be it a short word, or text that could fill a book, the hash generator will chew on it and return a fixed length.

If you make even the slightest change, the hash that is generated will be different. In the example above, I just removed the period appearing at the very end of the lyrics and this is the hash that generates:

92e627d6e77f819e15d78f8a1c938559

If you compare the two, nothing but the last period was removed, and yet the current hash generated is significantly different from the first. On the other hand, if you keep hashing the exact same content, you will get the exact same hash.

Bitcoin uses SHA256, which stands for "Secure Hash Algorithm 256 bit." The odds of penetrating it are improbable because the amount of computing power it would take to break it is more than all the energy available on earth at the moment. So until someone invents a way of computing with zero energy, cracking the SHA256 would be impossible.

As the book unfolds, you will see more features of the hash as you get a working and functional handle on hashes and mining.

Recap

Before we get any further, let's just put the whole thing in context so that we get a functional understanding of what cryptocurrencies are and what underlying elements and processes come into play.

Cryptocurrencies are cryptographic tokens of agreed value that can be sent between nodes in a transaction. There are no physical elements that are needed to represent these tokens, and they can be transmitted electronically. To objectify it, these clumps of text — called strings, are your coins. That's' the first part.

The second part of it is the blockchain, or the underlying records of transactions that give each coin its legitimacy. The block chain contains the record of the ledger that spans the life of the coin and the life of the entire cryptocurrency. Bitcoin is one cryptocurrency and the blockchain it runs on has the record of all transactions of each coin.

The third part consists of the nodes. Without the network of nodes, there is no credible record of the blockchain and this means that there is no reliable way to determine if a coin exists or is legitimate. The legitimacy of the record is because there is a record of it across independent nodes.

The last and final part of the cryptocurrency is the miner. We haven't talked about the miner yet even though it is the topic of this book. The miner is the node that generates the hashes which are placed in the blockchain. Without the miner, the hashes will not be computed and the blockchain will come to a halt.

Each node typically has a wallet within it and it really can have as many wallets as it likes, or it can have one wallet with as many accounts as it likes. In cryptos, they are not always called accounts, but rather, addresses. That's just like an email address, but in this case it is a string of characters.

Wallets

It's not that leather binding you keep in your back pocket or your handbag. A wallet is an app that helps you do a couple of things. First and foremost, it generates an address for you based on the rules of the crypto you are using. This is the address you give to your counterpart who is paying you.

Private Keys

In the process of getting an address, the wallet first generates a random number. This is the starting point. This random number follows the guidelines set by the wallet in accordance with the cryptocurrency it is supporting. If you take Bitcoin as an example, it is a number that falls anywhere between 1 and

1157920892373161954235709850086879078528375642790749043826051

It is represented in hexadecimal and that looks like this - between 0x1 and 0xFFFFFFFF FFFFFFFF FFFFFFFF FFFFFFFE BAAEDCE6 AF48A03B BFD25E8C D0364140.

Private keys are the most important part of all this and should be treated like you would the keys to your safe.

Public Keys

The public key is the next in the sequence of things that go into creating an address. This public key is derived from the private key, so in essence a random number in the range given above is taken and processed. The process is a mathematical operation that has to do with elliptic curves and is beyond the scope of this book. But that operation results in a string of characters.

Private Keys

This is an asymmetric relationship. Only one private key can exist for one public key, but 2E96 public keys can exist for one private key. So the only way to crack it is to try each key and see if it matches the public key that you can see. If you did this, there are 2E96 possible combinations, which works out to be 7.92 x 10E28 that you would have to try to be able to decipher the private key. This is the reason the public/private key pair is secure.

Bitcoin Address

Once the private key is generated, it is taken and processed again with the public key and processed in hash to result in a string of numbers. This is the Bitcoin address that you give to others to send you a payment.

Account Balance

Once you have an address and you have the private key, someone could send you a payment that would not appear in your wallet. But your wallet does not know it was sent to you because it was notified by the sender. It wasn't. Your wallet knows that the money has come to you because it is constantly reading the ledger that holds all the transactions in almost real time. When it sees there is a transaction that has your address in it, it shows you the amount in the account.

Once a particular address is in possession of a coin, the only way it can be spent is if the person spending it has the private key associated with the account. If they don't, the blockchain community will not accept the diversion of the funds and thus the block chain, or the ownership record, of the particular coin will not change. Each transaction that is in the record of the ledger really is just the recording of the ownership of a particular coin.

Blockchain

Now that you understand the transaction, we finally come to the blockchain.

Before that, let me back up to an earlier step. When we first looked at the nodes, we saw that they share information with other nodes. When you conduct one transaction, that transaction is sent out by broadcast to all the other nodes. When those nodes pick it up, they send it to other nodes, and at some point in time one of those nodes is going to be a miner's node.

A miner takes this transaction and puts it into a block. At the same time, other miners that get this pending transaction put it into a block as well. At any given point, there are multiple miners processing a given transaction as part of different blocks. Once they group a number of transactions together, they start to solve a puzzle that is resource-heavy. This means they take the hash of the last existing block in the chain and add it to a few different components, including the transactions they have chosen. They then try to solve a puzzle that is supposed to generate a hash within a certain parameter. This is actually much harder than it sounds and we will get into the details as we get further into the book.

Once the block puzzle is solved, the miner will broadcast to the rest of the network, and the first one to reach 51% of the existing nodes will validate their block, then fall off and attempt the next block. Whatever transactions that don't make this block will then be left as a pending transaction. The next miner to try the next block then pulls it and tries to solve the next puzzle.

Once the miner solves the puzzle and it is accepted across the network, he is rewarded with bitcoins for his work, then they're sent to his miner's Bitcoin address automatically by the system.

These bitcoins are brand new and have not been used before. The system mints these as a reward to the miners for their work in processing the blocks that go into the blockchain.

Chapter 2 Bitcoin Mining

Now that you have a cursory view of crypto coins and how they came into existence, what you want to know is how you could get into mining, and whether or not it's worth the time and investment you have to make in order to be able to mine the necessary coins.

There are two ways you can be part of the mining process. The first way is to create a bank of processors that solve the puzzle and then pass them into the general population in return for the reward. The second way is to become part of a pool that generates bitcoins and then share the reward for the world done.

There are a couple of parameters that you need to evaluate in order for you to be able to determine the profitability of the endeavor. Where you live, the cost of your electricity, the cost of the hardware that you need and your tech savviness all come into play. The goal of course, is to be able to program and get the hardware to efficiently mine the answers to the puzzles that reward you with the bitcoin.

We will go over each of the elements that you need to determine this in the upcoming chapters. For now we will look at Bitcoin mining, which is specific to Bitcoin and not necessarily the same with the other coins. There are some similarities across the different coins but that may not always be the case. However, the logic in determining them is still the same.

To understand the Bitcoin mining process let's look at the elements you need to bring together to be able to collect the reward.

What needs to be Solved?

We talked about this earlier, but it was a cursory visit. Now we will go into this in detail as it pertains to Bitcoin.

The Bitcoin algorithm gives you a puzzle to solve. You simply have to put all the different elements of a block together and then pass it through a hashing algorithm. That will give you a hash, which.0 needs to be within a certain parameter.

Once a miner groups together a number of transactions, they can choose whatever they want or just grab them at random. Then — as you already know — place them in a block. To solve the puzzle, they need to take a number and add it to the block and then hash it. What's the number? Well, it can be any number. Miners just keep guessing and trying until they get the outcome they want.

The number, by the way, is called a nonce. There is more than one possible nonce for the block and it may not always be the same nonce, because different miners will use different transactions to create a block. That will cause the combined nonce and block to result in a different hash. So what miners typically do, is try random number as the nonce until they get the right hash. The average time it takes for the system to create a block is around ten minutes.

But don't let that fool you. Just because it takes ten minutes doesn't mean you are going to be collecting the reward every ten minutes if you only own one piece of hardware. It does not work that way. It comes down to what is called a hash rate. We need to look at your hashing power to be able to determine how much power you need to have and how many times you can hash in a given period of time.

This may get confusing, so let's look at this again.

There is a significant amount of computing power in the expanse of the Bitcoin universe. At the time of writing, the entire Bitcoin universe is performing about twenty trillion hashes per second. That's a lot of hashing power. Remember, a hashing output is the output you get after running the algorithm and converting the number of transactions into blocks.

The blockchain is the heart of the coin, especially Bitcoin. It all comes back to the blockchain and for the blockchain to remain alive. If miners do not set up the transactions into blocks, then the entire things stops in its tracks.

But there is also the need to control the amount of Bitcoin that comes into the market. There is a cap as to the total number of bitcoins in circulation. That number is 21 million. Once it hits that number, no more bitcoins will be created, and the number of bitcoins will stay at 21 million or start decreasing over time.

How does it decrease? Well, I know this only too well. Back in the day when Bitcoin was only worth about $13, I had my wallet in a computer that got fried. I had about 30 coins in there and never bothered recovering them. I had not backed it up, and I hadn't papered the private key. How do it think I felt as it went to almost $20 grand? So anyway, those coins that I had on the drive and never used, are lost forever, not just to me, but to the community as a whole.

So there is a way that coins could decrease over time. But not to worry, the coins are divisible and we can create as many divisions as we need. But that comes later. For now the point is that miners only get paid in bitcoin until the time when there are no more bitcoins to mine. That date is fixed and so is the number of coins that come into the market in a given year. That number decreases every few years, and the way that is done is by reducing the reward for miners. When it first hit the market, miners were getting 50 BTC for the block. Soon, that dropped to 25, and now it's at 12.5 BTC for a block. Bitcoin is designed to generate a block every ten minutes and so 12.5 BTC enters the market at that rate. In a day, that works out to about 1800 new coin. That internal rate has been hard-coded into the system, and that's how much the computers run by the miners need to work at solving the block.

There are currently 16 million BTC in circulation. It's maximum circulation will be 21 million. That's just over 4 million left to go. If you are planning to go into mining, this is one point you are not going to hear anywhere online — so you better read up and pay attention!

There is a cap on how much you can mine, and that cap is something that is hard-coded into the software. The designers of the software calculated exactly how much a CPU could thrash out at a given moment and the upper-limit of processing it could handle. What they also did was write into the software how much computing power was spread across the mining universe. Then they would calculate the difficulty of the puzzle that was needed to solve the block. They adjust this till the output is ten minutes.

The interesting thing about this, is that the difficulty of the puzzles which the computers need to solve is controlled by the software of the currency. It is programmed to increase in difficulty as the computing power of the aggregate mining network increases. But in the end, it is still supposed to take ten minutes, so you can tell how much power you need.

Before we get into that, we need to recap some stuff about hash and hash rate. Remember that hashes are just the process to get a string of characters, whatever they may be and however long. Theoretically, you could take the city phone book and put it through a hash function and get a hash out of it. It doesn't matter what size it is, it needs to be run through an algorithm and it always comes out the same size, depending on the algorithm you use. So the resulting number is not a random string; it is extremely precise.

Everything that need to go into the block is prescribed. You have to have the hash of the last block, as well as all the transactions you are including. Also, you must have the nonce — remember that number we talked about.

Now let's look at the nonce in Bitcoin closely to get a better idea of what is going on.

Bitcoin tells you that the final hash is always going to be 256 bits long. What they will prescribe is how they want the hash to look. Since you can't control the rest of the information in the block, the only thing you can control is the nonce.

So in the block that goes in next, you enter the hash from the last block, the data that is to be included in this block and a number — the nonce. You get the nonce by guessing. Because what you want to do is get a specific kind of hash at the end, and Bitcoin will tell you if they want 30 zeros in front, or if they want 32, or 40. The number of zeros in front is a random occurrence because you are trying random nonces with your attempt.

That kind of trial and error is labor-intensive. The processor is trying hard to look for the nonce that, when added to the block, returns a hash that falls into the prescribed format, i.e. the number of zeros in front.

Once your processor gets that nonce, you send the whole block to the network and Bitcoin can easily check what is called your proof of work. Because there is no other way that you could have gotten the nonce without working it one at a time. Each time you run the block with the nonce, you are hashing it. So the faster you hash, the more you can convert and the faster you can get to the desired hash output.

There are a few ways for you to think about entering the mining market and make this profitable. Entering the mining business just because you can use your home PC or MAC and think you can make some spare cash is not the best way to do it, especially if you think of going into BTC mining. My advice is not to, and here is why.

Everything from this line onwards and to the end of the chapter is specific to BTC.

In the middle of 2010, the hash rate (that's the number of hashes that were calculated every second across the Bitcoin universe) was 1 billion hashes per second. What does that mean? That means, regardless of how many people running their computers and processors at whatever speed, all together they were trying for 1 billion hashes per second to get the block to come up with the hash that Bitcoin wanted. That would be the format with the 30 zeros in front, or something like that. Bitcoin controlled that format; it could tell you it wanted 10 zeros if it wanted to. I'll let you in on more about that in the next section.

That was in 2010.

Last week, at the time of writing this book (it's 2018), the universal hash rate was at 22 million Terahashes per second. That means 22 million trillion hashes per second. That is an enormous amount of processing power! It shows that the mining activity for Bitcoin is really high. But what it also means is that since Bitcoin wants to keep the time between new coins coming to the market at a steady ten minutes, the way to do it is to increase the complexity of the puzzle.

Remember we just talked about that. When the amount of computing power is so high, instead of speeding up how many coins it floods the market with, the algorithm demands a puzzle that has a lower probability of success. So it thereby requires a proof of work that is harder to get, and you have to end up doing more computations to get more hashes. In the end, it raises the difficulty so much, that the interval between reward events comes back to ten minutes.

At the moment, there is computing power of approximately 22 million trillion hashes per second. Per minute, that means 1,320 million trillion hashes calculated around the world are contributing that much computing power to generate that much work.

Here is how that calculation works out. If you go to the Bitcoin website you can see what the average hash rate is. The hash rate is across all miners in the network, although we do not know how many miners there are. We do know there are very few individual miners left. Most of them are either now professionally managed or pooled into large groups.

Here is why.

With 22 million trillion hashes per second, that works out to be 1.32×10^{25} hashes that need to be solved before getting the correct nonce to get the block reward. The reason we know this, is that Bitcoin has set the production rate to ten minutes. That means when you aggregate the amount of computing power, then that's the highest computing power that you are going to need to get to the solution.

There is a clear conclusion to be drawn here. First off, don't bother, if you are only using your home or office PC. It's not going to work.

Secondly, you need to ramp up your hash rate as much as possible to be able to get a periodic win. Let's assume that two players are all there are in the world to hash out solutions. The logic dictates that the two will have equal chance and equal outcomes over the long term. Both will attain 50% of the outcome. But the percentage of chance at the outcome is based on the comparison of the total computing power. In reality, this is the way it works. Your chance of hitting the block is a ratio of your hash rate against the total hash rate of the collective universe.

There are a few factors that return higher profits in the mining industry. The first is that you have a higher hashing rate. The highest hash rate that you can purchase at this point is the Antminer S9 and that runs about $2000 and returns 12.9 Terahashes per second. If you run your machines well and give them the necessary cooling they need, then they will last you for at least three years.

If you know that they are going to last you about three years and you are going to spend $2000 for them, then let's say you have your sunk costs all figured out. What you then have to consider is the cost of your electricity.

From there, you will be able to plug it into a calculator and figure out how much you will be able to make in a month. There are a number of good and powerful profitability calculators that you can use to see if you can make a consistent profit using the miner of your choice. You don't have to get the Antminer, but you do have to get something that is going to be able to consistently give you a high hash rate.

There are a few things you should keep in mind when it comes to the qualitative and quantitative aspects of making a decision to start mining. First of all you need to see what your budget is to invest in the hardware. Once you do that, you should look at what equipment you will need. For instance, let's say that I have $20,000 to invest in a Bitcoin mining operation. That will allow me to get about nine Antminer S9 miners, and the balance would go towards the other stuff like a rack, fans and cables. Once you have that, you need a place to put all the equipment. if you only have nine miners, then a simple rack in your room will do — but make sure you have sufficient airflow in your room and enough ability to cool it. If you do not keep your cooling in check, you will very likely burn out the equipment, and that can get very expensive very quickly.

Once you have that done, grab your monthly electric bill and look at what your utility is charging you per kWh. You need this information because mining operations are energy-intensive.

One you have that information, go over to an online mining calculator to look at your monthly profitability predictions. Now what pieces of information does the calculator need from you?

The first thing you need is the difficulty rating. For that you can go to Google and type in "current Bitcoin difficulty rate." I just checked for this example and this is the current difficulty level:1,590,896,927,258. You need this number to plug into the calculator.

Once you plug this in the calculator, you'll be required to find the hash rate of your miner. If you only have one miner, then use the hash rate for that one miner. But in this example you have nine, so if your hash rate for the one Antminer is 12.9 TH/s, then for nine of them it's going to be 116.1 TH/s. Plug that in.

You will also be asked for the power rating of your mining rig. All you have to do is check the power rating for it and multiply that by the number of rigs you are running. In this case, my sample miner uses 1,375 watts per miner. Since I have 9, that's 12,375 Watts. You also need to find out the cost of electricity. Mine here in California is 12 cents.

Finally, you need to plug in the cost of the equipment. Once I plug all that in, I also need to plug in the exchange rate for BTC and USD. At the moment it is just north of $8,000 per BTC. With all that plugged into the calculator, the result I get is, in a month I will make 0.535 BTC. Also, my electricity will cost me just over $1,000. In that month I will make just over $3,400.

So if I make $3,400, I will be able to break even, assuming the price of BTC to dollar remains constant in six months. After six months, then I start becoming profitable. At this exchange rate, my profit for the year would be 100% in the first year, 200% in the second year, and again, 200% in the third year. This is assuming there is no increase during these three years in the cost of electricity, the difficulty rate or the number of miners.

Of course we cannot be sure of any of that. But there are a few rules of thumb for you to follow.

The first is, if the exchange rate increases, the number of people coming in to mine will increase and the difficulty will go up. If you want to keep your same ratios then you would have to make a corresponding increase in your own investment, but that is not always possible in these situations. You have two choices. You can move the operations to a less expensive state where you can save on electricity. In the last example, I used 12 cents. If that were halved, my profit would go to $3,960.

The second choice is that the cost of the rig makes a significant impact on the investment. The best way that you can look at it is to see the ratio of hash to price. So for instance, the Antminer S9 runs $2000 and gets 12 TH/s so that would mean I get 6 billion hashes per second per dollar. If you take all your options and divide it down to the hashes per second per dollar, then you will be able to get a better picture.

That's one way, but there is an even better way, especially if you can compare them based on the profitability based on wattage. Then divide the last number you got — 6 billion hashes/sec/dollar divided by the total wattage of your rig. In my case, that was 1375, which means my ending number would look like this: 4,363,636 hashes per sec per dollar per watt.

That number by itself does not mean anything at all. However, for it to make any sense you would have to compare that with other rigs and other configurations. So let's say you look at the Antminer R4. In this case the R4 costs about $1000, which is half the price. I'll put $20,000 into it again and so let's see how much I get in return when I compare apples to apples.

In this case, with all the same inputs, I invest $20,000 and spend $18,000 on the hardware to buy 18 individual pieces, and keep $2,000 for ancillary expenses. What I am left with is a rig that has 18 miners, and according to the calculator I make $4,800 per month. This is a slightly better profit rate than the earlier one with the S9 since I only made about $,3400. That's quite a big difference as far as profit potential.

You can also look at it in the second way I talked about, with 8 billion hashes per sec per dollar. You get that number by taking the hash rate divided by the price, minus 8.6 TH/s divided by 1,000 dollars. This gives you 8.6 billion TH/s/dollar. Now divide that by 845 watts. That's 10,177,514 hashes per second per dollar per watt.

In this case, there is a clear winner and that is the R4. It gets you higher hashing power compared to the S9 when you stack a few of them. If you are just going to run one, then the S9 is simple enough to work with.

Mining Strategy

There are a couple of different mining strategies that you could engage in when you decide to start mining BTC. The first thing you want to do is go in all the way. Do not enter this market if you just want to put in 2 or 3 thousand dollars and just get your toes wet. That's a waste of time. You will lose money and you will hate the industry and yourself for not making it. You'll be better off taking the guys out for a beer and spending that money!

When you want to play this game, remember the old adage of strength in numbers. Go in with a couple of friends and start off with no less than $100,000, then set up the system in the coolest room in the building.

What most people do not understand is that it is not just about the number of fans to make sure that the processes do not overheat, it is also about keeping them at their optimal temperature. Overheating is a major problem, but optimal temperatures increase the average hash rate.

In the winter, open up the windows if you have to, and in the summer crank up the air and keep the fans channeling air out of the room. If you have an attic fan, that will work great to keep the airflow in the home or office better regulated.

The next thing you need to do is have standby fans and power generators. The power generators are not for the processors to keep running if the power goes down. They are there to keep the fans running in case of an outage. If you do not run the fans and the processors suddenly stop, they are still producing heat and now the fans are off too — that's a problem. So make sure you have back-up power to run the fans and to gently wind them down after a power disruption.

Run the machines 24/7. Running them at shorter intervals doesn't increase the life of your equipment; keeping them cool does.

You are going to enter the game. This is a business just like any other. The time to think of bitcoin mining as a hobby or as a pastime is long gone. Bitcoin mining is a serious business that requires that you are willing to understand the idea of cryptocurrencies, the fundamentals of cryptography as it pertains to coins, and the operating systems and software that are involved in getting this hardware to run at the most optimal rate you can devise. There are numbers of tweaks and add-ons that you can take advantage of to make your rig perform as efficiently as you could hope for.

Deciding to Mine Bitcoin

At the end of the day you need to focus on two questions. Firstly, do you want to mine cryptocurrencies? The idea of mining cryptocurrencies should not just be something that you mistakenly think you can jump into without forethought and planning. It is, after all, just like any other business, with an investment requirement, the potential of risk and profit, and the need to apply consistent effort and problem solving. Not to mention the need to constantly upgrade your skill level. It doesn't mean that if you can turn on a computer, that you can make money mining. Even if you join a pool, it still requires some sort of knowledge of what you are doing so that you can counter any obstacles along the way.

Once you have decided to mine coins — and the reason could be one of many — the next thing you need to realize is that it all boils down to one factor.

Let me explain.

You will read across the web that there are multiple factors that go into evaluating the infrastructure and the mining power. Yes, those need to be done when you decide which is the right miner to purchase. But those are not the same considerations you need to make when you decide which coin to take on. There are minor tweaks that you need to make that work best when you are an expert at one. The learning curve may not be steep but it does exist, and as time goes by, being an expert at one will be far more beneficial to you than being average at a few.

Putting aside hardware and energy costs, as well as time to divert to the project and start-up costs, there is only one thing you need to study and understand. That is the market price of bitcoin. Why? Because unless you live in an area where bitcoin has penetrated every area of your life —you'll still have to buy your groceries, get your hair cut, pay for gas, utilities, cable and even your kid's lunch — by converting that bitcoin to legal tender. You know, into dollars or whatever your native currency is.

In that case, you are going to need to keep an eye on the exchange rate. Think of it this way. If you bought a plot of land that is supposed to have gold deposits, then you need to see what the price of gold is on the market. If you think that the price of gold is on its way up, then you know that the profits in the future are going to be something that will make it all worthwhile.

That is the key factor — where you think the price of bitcoin is headed in the near and long term.

That is something that you have to ask yourself. It is the strategic question. It doesn't matter how much effort you put in, if the price of bitcoin is stagnant or depreciating, you are going to find it less attractive, and that is something you need to decide now. There is no right answer, but once you make the decision, you will have to live with it.

You also need to decide what currency you are going to settle on, because the hardware, settings and learning curve are all different for each. So as I have said before, you should be an expert at the one you choose, and that expertise comes with experience.

Chapter 3 Ethereum Mining

Each coin is mined differently and each coin obviously has a different reward scheme. On top of that, the market price for each coin varies vastly from a few cents to tens of thousands of dollars, so you have a very diverse landscape of reward profiles.

In the last chapter we focused on Bitcoin. In the rest of the book we are going to talk about the top three altcoins and how and why you should mine them. Admittedly, if you are new to mining, BTC may not be the place for you to start.

There are other coins that offer higher coin returns and lower block times, but the rewards and market prices are lower. The thing about the altcoins that we will be looking at is that they are all used as currency to purchase whatever you intend on purchasing if the vendor allows it. It's like Bitcoin, but just a different coin. We are only looking at these at the moment in terms of mining. There are other coins and they were not chosen to be included in this book for one of two reasons. The first is that they are too small to be considered profitable for you just yet. Reason two is because they are not financial in nature — they are more like a token that is used to purchase or exchange proprietary goods and services.

What are altcoins? Simple. Any cryptographic coin that is not a bitcoin is an altcoin. There are more than 1200 altcoins in the market now. Some of them are proprietary, meaning that they're used within a closed group. Others are public and can be used in an open group, so anyone can buy, sell, trade, mine them — anything, really.

We will concentrate on the ones that are open to all.

You have already seen how to mine and what to do when it comes to Bitcoin. Now we will look at Ethereum, Monero, and Zcash as representative of the altcoin universe, and whether you can find a coin that is worth it for you to invest in as a miner.

Ethereum

The point you should be familiar with by this time is that mining is not about digging anything! It is about creating blocks, which is especially true of Bitcoin. All you are doing is arranging the information into the ledger and getting them to be accepted into the chain of events, or the chain of ledger entries. Each block has thousands of entries that are stitched together. Why do this? Because the system is decentralized and the best way to get someone to do it, is to be able to have proof of work to determine who gets to put the proper transaction into the block.

The main difference we talked about in this book between mining Ethereum and Bitcoin, is that in Bitcoin we use specialized hardware like the ASIC, then use Antminer to be able to mine the blocks. In this section I will take the opportunity to discuss how you can mine Ethereum using graphic cards, without the need to get prebuilt hardware like we did with Bitcoin.

The first thing you need to do when you get into Ethereum mining, and pretty much for any other coin as well, is to get a wallet. Wallets are free and the wallet gives you the ability to receive coins from others — or in the age of mining — directly from the system itself. You need this wallet so that you can receive, store and spend your coin.

So the first thing you need to do is go to http://www.myetherwallet.com.

Once you get there they are going to ask you for a password. Set your password up and make sure you use a really strong one. The way I create passwords is to go to a random password generator. You can use http://randompasswordgenerator.com.

This will give you a strong password in a format that you chose. I suggest using upper and lowercase, numbers, and symbols. It should look something like this:

y\e@uXY5\nV3FfcF

Once you have your password, before you put it into the wallet setup, place it in a text file and run that file through an encryption algorithm. I typically use https://codebeautify.org/encrypt-decrypt.

Choose whichever form of encryption you like from the many out there, then provide a key (a simple word that you can easily recall). And that will encrypt it. This is the encryption I used for the above password:

THgl/nwIHGsyumIbsWeh6HurOVAx0RWS+xhHnL5N+Ux4o1ARznuv/5J/6Hw3CMaqUn6AJJL

For the fun of it, if you took this encrypted password to codebeutify.com and *decrypted* it, you wouldn't know which one to use. But if I told you that I used Rijndael-256 in CBC mode and my key was 'bike,' then you could plug that in and get the proper password. This way, you will significantly reduce the chance of your password being compromised.

Once you have your password set up, click ok, and it will take you to a page that you'll need to store your KeyStore file. This will give your computer the encrypted connection that it needs when you are accessing your account. To do this, I suggest you set a brand new thumb drive and save the file from your computer directly to it. That way when you are done, you can remove the drive. Then if you are ever compromised, the information in the KeyStore file will never be available on your computer as it will be effectively air-gapped and offline.

There are three things you need to do to this file and thumb drive. First, you have to make sure that you do not lose it, because that is the only way you are going to be able to have access to your wallet. The second is never share that with anyone. Finally, make a backup of it. I have a backup drive that I put everything into and leave in my safe. You can do that as well.

The next step is to create and save your private key. Remember in the Bitcoin chapter before this we talked about the private key. This is the private key that your account is based on and so you need to keep this private key safe because it is the private key that releases the use of funds and it is the private key that is proof of funds as well. The only way you can stake a claim to the funds is by the ownership and possession of the private key. Make sure you do not lose this. If you do, all the coins in the account will be lost forever, unavailable to you or anyone else.

Once you have set up your account, it is time to get together with a pool administrator to get yourself into pool mining.

For this, go to nanopool.org.

Here you will be presented with six alternative coins to be able to mine. For now, choose Ethereum. Not Ethereum classic — that's different. Click on the button that reads, "Quickstart." Click on that and follow the instructions. It will take you to a page to optimize your system and then it will give you an opportunity to download the software. In there, you would have to input your email address and the Ethereum address. Once you do that you are ready to go.

If you notice, this is pool mining. We will talk about the benefits of pool mining and also about pool mining versus stand-alone mining. This will show the benefits of the two alternatives when we hit Chapter 5.

Understanding Ethereum

Unlike Bitcoin, Ethereum is not solely a currency, and any notion that it is just like Bitcoin and only used as a method of exchange is false. Ethereum, even though it has a similar blockchain, coins, and rewards, is really a very large ecosystem of computing power. In other words, it is a decentralized platform of intelligent system platforms. You can also think of it as a platform for other platforms.

But that is not our scope at this point. Right now, we are only focusing on its mining aspect, and the rewards we get for that mining are Ethereum coins.

Ethereum also works on top of a blockchain called the Ethereum Frontier. This is much like Bitcoin, except it is not the same blockchain, so it has some minute variances in its execution. If you have the time, hope on over and take a look at the manual which you can find here: https://ethereum.gitbooks.io/frontier-guide

As a miner, perusing the manual is not entirely important, but can be if you want to take it to the next level to build products on top of that. For now, we won't worry about the nitty-gritty details outside the mining areas of the protocol.

The blockchain also uses an incentive base, just as Bitcoin does, in as far as having miners create the block that need to be formed into a chain. The work that they do is then presented as a proof of work and submitted with then block they are submitting and that is verified by the rest of the network participants. This is the typical blockchain sequence of events.

In the same way that Bitcoin keeps the new coin creation to ten minutes by controlling the difficulty of the puzzle, Ethereum also follows this "block time." In Ethereum however, block times are much lower.

How do they control block times?

Easy. There is a direct correlation among three variables: difficulty, hash rate, and time. As hash rate increases, and difficulty remains the same, then time will decrease proportionally. Why? Because the more you can hash, the more numbers you can try until you get the nonce that works. If you increase the difficulty and there is no change in hash rate, then time changes proportionally to difficulty.

How do you change difficulty? If you remember, in the Bitcoin section we talked about changing the requirement for the hash. We will look at this again in a little more detail. Ethereum uses the KECCAK-256 hash. When you get a 256-bit hash of the entire block, it looks something like this:

3061b05cebb8e9b71be795bdf59c014983dd654b014db0a662f2a9826bd0a957

That string of characters is the hash of the entire Preface section of this book, containing 378 words., or 2104 characters. It was processed and altered to the 64 characters you see above.

In Ethereum, that is called the partial proof of work, and it is not enough to claim the bounty or the reward. Instead what is needed is a hash that is of a particular configuration. And that is the puzzle — you need to figure out what that configuration is! If I took that same Preface section and now added a number that goes between 1, to let's say, 40,000,0000. You can follow along and try this for yourself. Simply go to an online hash generator like the one here:

https://emn178.github.io/online-tools/keccak_256.html

And you can copy and paste any text to give you the keccak256 hash (remember this is the one that Ethereum uses; Bitcoin uses a different one). Once you get the hash, now go and find a hash that starts with the number 2. Since you can't reverse engineer it or reverse calculate it (remember it is asymmetric), you will have to guess your answer. So let's try this:

I add "1" at the end of the content and I get a hash that is something other than the one I want (starting with 2). So I keep trying numbers until I get to a point that the hash starts with 2. In my attempt, that number was 12114.

So now I submit my proof of work as that number 12114, and the network checks it and agrees that the nonce of 12114 indeed does result in a hash that starts with 2. That was fairly easy and it took me a few seconds to get the answer.

Now if the mining world started to increase its hashing power and the network notices that the block times are decreasing, what that would do is change the puzzle. Let's say instead of just 2, it wants a nonce that will result in a hash that has 10 zeros in the beginning, or 20 zeros. Whatever — it knows that by changing the difficulty, then the time interval changes.

Since there is a direct correlation between difficulty, time intervals, and hash rate, the more miners that come in to hash the blocks, the difficulty of the puzzles increase to compensate for it and keep the intervals constant throughout.

Difference Between Blockchain and Ethereum Frontier

From a functional perspective, there is very difference between Bitcoin and Ethereum from the perspective of mining and mining algorithm. Aside from the fact that they use different algorithms, the functionalities are the same.

The one outstanding difference is the time intervals between the blocks, and that makes the computations to get the proof of work easier when you are mining for Ethereum. That ease of operation means that there isn't so much required in terms of hardware for the mining of Ethereum versus Bitcoin.

For Bitcoin, you will undoubtedly need specialized equipment called ASICs. ASICs are "Application Specific Integrated Circuits." They are faster, consume less power and are designed to hash. They cost anywhere from a thousand to five thousand dollars per unit and have a fairly low price per hash ratio.

On the other hand, because you do not need such high-level processes when mining for Ethereum, you do not need such high-end machines, either. All you need are graphics cards that have GPUs. GPUs are "Graphic Processing Units." They are like CPUs but are more powerful because they were designed to process graphics at a faster and smoother rate. If you have a powerful gaming machine, then you will have no problem converting it into an Ethereum mining computer.

There are three ways you can get into Ethereum mining.

1. The first is to become part of a pool as we talked about in the earlier part of this chapter.

2. The second is to get your computer to work in the background and mine Ethereum.

3. The third way is to make a go of Ethereum mining as a business, and this is where we will show you how to make money on your venture.

The first thing you want to understand is that the market price in dollars for Ethereum is $880. Your first question is to think about where the coin is going in the next one, three and five years. It is an educated guess at best but if you are not optimistic about the price of the coin then you shouldn't be in this market. The way I calculate the reasons to be in the market is never on the price of Ethereum at the present moment. It's about what I think it will be in the future. This is how you have to think about Ethereum and why you should want to mine it.

Let's say you project Ether to be $3,000 in the next year. Now remember, I am not telling you where Ether is going to be a year from now; that is for you to determine based on research and expert opinion. I am merely using that figure to show you the train of thought that flows from the number you pick.

Once you have determined your projections for the end of year one, and those for the end of year three and five, then you can start to look at the investment strategy you'll deploy that is risk-adjusted and takes advantage of the best returns possible.

Here is how you do that:

Your first option is to buy the Ether now and spend $800 per Ether doing it. Let's base this on a hypothetical $20,000 investment. That amount of investment will get you 25 Ether. At the end of Year 1 with your price target being at $3,000, 25 Ether would be worth $75,000. That is a whopping $55,000 profit, or 275% return.

Now the point to remember is that I am not advocating or suggesting that Ether will return this kind of price in a year. I am merely showing you the train of thought that you should have in evaluating whether you should buy or mine. The next part of the example gives you the comparison you should need.

Let's say that your year three target is $8,000 and your year five target is $12,000 per Ether. That means that you will have a return (uncompounded) of $200,000 in Year 3, or 1000% return or 333% return per annum. And in year five you would have $280,000 return, or 2400% or 480% per annum. Pretty fantastical stuff, but remember this is only for illustration.

Now let us compare this to mining.

If you start with a $20,000 investment, and go into mining and pick up a fairly good rig, this is what you will be able to achieve.

I will give you a couple of setups on the mining rigs that you can build with this, but there are three very important things that I want you to understand before we get into it.

The first and most important thing you should keep in mind when you want to get into Ethereum mining is that there is a good chance that Ethereum stops the mining process sometime this year (2018) by stopping the Proof of Work part of the infrastructure.

If that happens, then you will not be able to mine Ether again. They may move from Proof of Work to Proof of Stake which is a whole different method. With the Proof of Stake method, you can only mine the amount of coins you already own. So the thing you need to know is whether or not you want to spend your time working on a rig to mine Ethereum, if there is the chance that they move from POW to POS.

The second thing you have to know is that all the GPU cards have been jacked up in price and in some places you can't even get them. The one place that you may be able to get them is on AliExpress, and they are typically cheaper there. But you have to be sure that you know who you are buying from, since they will be a no-name brand but typically have the same specs as the more expensive ones here in the US.

The final thing you have to understand is that you can mine a number of different coins and not just Ether when you use the graphics card, so even if you are no longer able to mine for Ether, there are other coins that you can move on to. If you go to pool mining with your rig, you will still be able to mine Zcash and Monero, among others.

So let's get started!

Get the NVIDIA GE Force 1080; it retails for anywhere from $800 to $950 per piece and it is able to clock in 31.2 Megahashes per second. If you put 20 of these together at $800 each, that works out to $16,000. You need to use the rest of the $20,000 (which we explained at the beginning of this comparison) on infrastructure, racks, fans and cables. It will all come up to about $20,000, and now you are on equal footing to buying the 25 Ether that we saw earlier.

Whether you hook it up to a mining pool or if you put it together yourself, you will be able to generate 19 Ether per year with this setup. I doesn't matter what the price of Ether is now, because remember you were looking at it from the perspective of an investment and that we do not need to liquidate it yet. We hold it for the time of one year, three years or five years as we did in the buy-and-hold strategy above.

In this case, if you do not increase the investment into the equipment and just leave it as the initial 20,000, here is what you need to keep in mind. Your electricity bill which was about 12 cents per kWh, works out to about $9,000 in the first year. You have made 19 Ether and so you need to sell about 11 Ether at $800 each to cover the electricity cost. That now leaves you with 8 Ether for the first year. In the second year, you have 16 Ether. At the end of the third year you have 24, end of 4th you have 32, and at the end of year 5 you have 40 Ether.

Let's now look at the decision to mine or buy. If you bought the Ether you would have 25 Ether at the end of the first, third and fifth years. That would not change, and you would be relying only on capital appreciation to make up your profit.

However, if you mine the Ether, then after paying your cost and keeping the rest as Ether, you end up with 40 Ether at the end of five years. That works out to be $480,000 instead of $300,000. You're ahead of the game by $180,000 if the price trajectory is as you thought it would be in five years.

But let's also look at it in another way to show that mining is a better alternative. The price of Ether stays stagnant at $800. Your investment goes nowhere if you bought 25 of them at 800. That means that you have actually not made any money in the nominal sense and have actually lost money if you adjust for inflation. But if you mined it, then even if the price of Ether stayed at $800, you would still make a profit of 15 Ether. That's because you would have 40 if you mined it and 25 if you bought it.

When you mine, this works for all the currencies, not just Ether. It works under the same logic for Bitcoin as well as the other altcoins.

As a miner, you need to be very shrewd with where you are going to apply your resources, and then take advantage of the market conditions to make your profit. If you decide to do Bitcoin, you will not be able to do Ether because ASICs do not lend themselves to mining Ether. On the other hand, graphic cards are not optimized to mine Bitcoin. So right there you see that you need to specialize in one or the other — and you should — so that you master one coin and then apply your resources to it.

The other thing you have to know about mining Ether is that it is not only processing-intensive, it is also memory intensive. So if you don't have at least 4GB on the card, you are going to have problems down the road.

Then again — and I repeat this — there is a good chance that Ether switches to POS from POW. This means the mining of Ether will only be done by those who already own it and if you wanted to continue to mine more, you would have to own more. You do that by increasing your stake with a purchase. If indeed Ether decides to go that route, you will see a hike in market price. For those who are really into mining, this is actually a great way to increase your exposure and investment with Ether. You see, it has a significant upside due the use of this platform as not only a currency, but an enabling technology for smart contracts.

Chapter 4 Monero Mining

The next in the list of possible mining opportunities is Monero. So far you have seen two kinds of mining. The first was done with specialty hardware — the ASICs for Bitcoin. Then you saw how you could use the GPU on a graphics card to mine Ether. In this one, you are going to see how you can use the CPU in your regular desktop to mine Monero. This means that almost anyone who has a computer can participate in mining activities!

You don't have to be a mining expert, or invest in specialty machines to become a Monero miner. It also means that you have a better cost control over the way you can mine on a large scale because it is possible to get old computers with CPUs on them that still work. Then you string these in a line and have them mine Monero. These can come pretty cheaply if you know what you are doing.

Monero is currently trading at about $200 per coin with a market cap of about $2 billion. It is pretty new to the market and pretty easy to mine.

Depending on what hardware setup you have, there are different software packages that you can use to set up your computer or your rig as a mining station.

Because you can use your CPU to mine Monero, there are a few things I have to warn you about since the hardware you are using may not be able to handle the load that is placed on it while hashing.

There are no moving parts in a CPU, but it is highly susceptible to heat. If the CPU overheats, you could irreparably damage it and that is the most expensive part of your computer. While it may be ok to buy a new CPU to replace it, you don't know if there is also damage done to the motherboard. The best way to prevent all this is to make sure that you do not overload the processor and that it is adequately cooled.

I have seen people leave their computer in the garage in the winter for this and run the cold night air directly to the fan inlet, and that keeps things nice and cool. But for the rest of the year, you need to keep the computer cool by blowing it with fans and keeping the air circulation flowing with the air outside. If it is summer and you live in a fairly warm place, make sure you have an air conditioner running to keep the ambient temperature in the room below normal.

If you are bringing in air from the outside, make sure you run it through a filter so that the dust in the air is removed before it is sent to the CPU fan. If there is dust in the air, then in time the accumulation of dust on the systems will complicate the problems of overheating as well.

I cannot stress this enough — the heat from the processor can destroy the CPU and the motherboard if it is not sufficiently cooled. You should also keep the area clean and dust-free.

Now that you have those things in mind, let me just add that Monero can be mined with the CPU and the GPU as well. So you can use a rack of GPUs in the same way you used it in the example to mine Ether in the last chapter. The GPU usage is interchangeable, and the difference in choosing whether to use the miner to mine Monero or to buy it as an investment would again depend on a number of factors. But it will also depend on your projection of the exchange rate of Monero.

The important thing I want to mention is that if you are indeed using just your computer's CPU and you do not have any other computational power, it is best that you do not try to do the operation on your own. There is only a miniscule chance that you will ever get any coins.

Here is why.

When you mine coin, the thing that is most important to remember is the hash rate. If you take the hash rate and look at the amount of aggregate hash rate in the total mining universe, you get an idea of the scale of your input versus the scale of the total market.

Now consider two things. The first is that the difficulty is adjusted based on the total hash rate of the industry. If you are just one GPU hashing at 1000MH/s, and the industry is doing 20 million Terahashes per second, you are a tiny fraction of that. But the difficulty is based on the total market. In this case, the group with the greatest combined hash rate is going to have the best chances of getting to the block first.

Let's calculate that out, assuming that the numbers are so large that the average hash rate corresponds directly to the block time. The aggregate hash rate of the Monero market is approximately 1 GH/s at this point in time (that will change in the future, but we shall use it for sample calculations here to give you an understanding of the way to use it for intelligence). That 1GH/s means all the miners put together are hashing out at a rate of 1GH every one second.

In a twenty-four hour period, that works out to be 43,200 GH per day. Or more importantly, 60 Gigahashes per minute. Correspondingly, the block time for Monero is two minutes. That means they have adjusted the difficulty of the puzzle to correspond to the POW to be found every two minutes. That means you need 120 Gigahashes to get to the solution. Do you see how that works?

If the industry is running 1GH/s and the block is coming out every two minutes that means it needs to process 120 Gigahashes. If you have a GPU or a CPU curing out 30 Megahashes per second (and that is high for a single CPU) it will take you on average 120 Giga / 30 Mega. This is 120,000,000,000 / 30,000,000 = 4,000,000 seconds. It will take your GPU (the 1080 that we used in the Ether example above) 4 million seconds to calculate that many hashes. That is approximately forty-six days to get one block.

But you have to realize this is a race; someone is already calculating it while you are. If you don't get there in time, they get the prize and you have to start from zero again. If you do not have the processing power, don't bother. You will just be wasting electricity.

This is the reason you join a pool. Because in a pool, the pool administrator divides the nonce range among all the computers that are logged on, and constantly distributes the calculations to allow the CPU to share work in its favor.

Think about it this way. Imagine if you had a possible nonce range of 100,000,000. That means it goes from 0,1,2,3,...1000,000,000. To get to the target nonce which digests the hash down to what the Monero network is expecting, it starts with one hash at a time — each time a successively higher nonce. If we did it that way, then take one CPU to calculate upwards, one at a time, you can see how long that could take! But suppose you had ten computers all with the same hash rate. Let's say you needed to chew 12 billion hashes to get to your destination, and you had ten processes running at 12 million hashes per second. You would get the answer within one-hundred seconds, which is just about two minutes (that is strictly coincidental — was not planning on that to coincide with the block time).

So the point is that when you are part of a pool, all of you get to the answer faster and you all divide the bounty. Plus of course, there is a percentage for the pool administrator.

For small investments, you will be better off going with a pool administrator. It only makes sense for you to go with your own set up if you are one of these mega-large mining farms that have thousands of dedicated ASICs or GPUs running 24/7.

With Monero, it is ok to mine with a pool — or even outside of one — as long as you get the hang of what you're doing. In fact, it is better for you to be able to start off small and keep mining within a pool, saving your coin. Then you'll get to a point where you can plough your investment back into larger and more sophisticated equipment. From there you can then create mining farms and go out on your own to mine for yourself.

Chapter 5 Solo Mining or Pool Mining

There are some significant considerations you must make before you decide whether it makes sense for you to mine on your own or join a pool. Actually, there are three that you need to think about before deciding.

The first of course, is if you have the funds to make it on your own. You are better off when you are in a group and you have significant bandwidth and processing power to be able to race to the finish line. If you don't, and think you'll still get some coins here and there, it's more likely that you will get none and your investment will turn belly up in short order.

Don't forget you are paying for an elevated utility bill by running your rigs (also note that the electricity bill that you get will be a little higher than what you calculate in the profitability calculators). Why? Because those calculators only take into account the power consumption rate of your card or ASIC. They do not yet count the lights, fans and air that you channel to keep the environment within the temperature range for the thing to work. If you skimp on that, there's a very good chance you will be ruining cards and having to purchase more, which increases your total investment and reduces your returns. Never skimp on cooling.

The second thing that you have to consider is how much your hash rate really is. If your hash rate works out that you are only going to be solving the puzzle every 2 months, you're going to be very far behind the curve, so you need to be able to get the hash rate that is enough to at least challenge the market at par. That is the rate we calculated in the last chapter.

Finally, there is the kind of pool that you are pooling with. If it is a proportional pool, then you will get paid by the amount of work you did in proportion to the total work that was done to mine that block. If your entire group is successful, then you get paid a proportion of the work. If the pool does not make the reward, then you get nothing. Read the fine print!

Then there is the second kind, where you are paid no matter what happens as long as you provided computing power. If you provided the power and the pool didn't get the block, it's ok. The pool still pays you for the work done.

They figure that it's better to keep you in the pool, pay you, and absorb the cost.

Just to recap, you need to be able to get the average of ten minutes per block out of your setup for Bitcoin, and two minutes for Monero, if you are going to try and go out on your own. If you do not meet this simple threshold, do not bother, because there is one other point I need to highlight to make sure you understand this.

Up till now, you have been looking at these profitability calculators, and the biggest misconception that everyone has about them is that they apply to times when you are on your own. They don't . When you are in a pool, the pool doesn't pay you for what you solve, they pay you for what you calculate. So as long as you keep churning and they see that you are churning, they pay you for the amount you hashed per second for however much time you were online.

This happens regardless of whether you actually solved any blocks or whether they did. But when you are on your own, remember it is a race. If someone finds the block before you, you get nothing. And if your hash rate is causing you to not get the solution within ten minutes, you are not going to get anything. All that electricity that you expended running the GPU or the ASIC rigs is all going to be your responsibility.

On the other hand, there is another way you could turn to mining, and that is for you to bring in equity partners and invest in a mining farm. How large that farm would need to be would be a function of the price of the coin, the price of electricity, the price of the hardware — and the most important one of all — the aggregate hashing rate across the market.

Here is how that will work. Let's take Monero, for instance. If you look at the hash rate, it is 1GH/s right now, and we know that the block rate is ten minutes. That means if you have the hashing power of 1GH/s right now, you would be able to control 50% of the mining industry in Monero. Let's look at that calculation here which will help you make your decision. That would be determining whether to mine a coin and to see how much you should invest, as well as the decision to be a part of a pool or to be on your own.

Solo or Pooling Calculation

You know that in Monero you need to solve an average of 120 billion hashes to get the answer. For you to get that kind of firepower let's look at the GPUs you need to invest in.

If you take the same GPU as the one we tried in the last example, the NVIDIA Ge 1080 Ti, then you know you are going to be able to do 600 hashes per second (the rate differs if you mine Ether vs Monero). How many units would you need to be able to do 1 billion hashes per second? Well, you would need just 1.6 million units of the 1080s. Each unit costs about $900 (more or less) so that means you would need to spend $1.5 billion to be able to get the hash rate that you need to halve the total hashing power equivalent to the current hashing rate.

But here is what happens If you add your 1.6 million units and your 1 GH/s capacity. The total hashing capacity of the entire universe doubles, and that means the difficulty would increase and it would require more hashing than you have. But that is ok. What it also means is that you have 50% of the hashing power of the total market , so you will get 50% of all the coins minted in a day.

If each block is every two minutes and the block reward is 5.23, then what you have if you get 50% of all the block in a day is 5,23*.5*30 per hour x 24 hours = 1882.8 XMR per day. In a month that works out to be 56484 XMR. At present the price of XMR is $250 per coin, which means on a monthly basis the reward is $14 million.

But from here you will also have to compute the rental for the space, the cost to cool the entire facility and the cost of electricity to run the GPU cards. To run the GPU cards alone at 12 cents per kWh would be $2000. If you can do it at that rate, then it becomes totally profitable for you to walk in and control the game.

It gives you 50% of hashing power and the rate of income would be easy for you to be able to upscale in the event other entrants penetrate the market. If you follow the logic of this calculation, you can start to see whenever it makes sense for you to do that, and which point you should enter or exit the market.

But you will need to spend $1.5 billion to make that $14 million a month, and I am guessing most of us don't have $1.5 billion lying around in the sock drawer. But you get the point. Unless you can put up that kind of firepower, your best bet is to go with a mining pool.

Pooling

Here are some of the pools that you can join when you want to set up your mining. For those of you who already have installed capacity and you want your extra capacity to be put to some good use, this is a good way to make some money on the side. It is not a full-fledged investment because you will not have to spend any money in doing it, except for the necessary electricity. But if you use the calculator, you will be able to see if it is profitable and you can make some pocket money with it. If you indeed are starting this way and starting small, it is best that you leave the coin you earn in the account, or transfer it to your wallet. Best not to transfer it to the exchange to convert it to your fiat currency of choice. That way you will be able to appreciate the appreciation of value over time.

MineXMR

1% Pool Fee

0.5XMR Minimum Payout

Moneropool

0.5XMR Minimum Payout

Nanopool

1% Pool Fee, + Payout Commission of 0.015XMR

1XMR Minimum Payout

Dwarfpool

1.5% Pool Fee + Payout Commission of 0.014XMR (exchange wallets) 0.008XMR (normal wallets)

You can find these pools with just a Google search. Make sure you find one that is the cheapest and also one that is the closest to you in terms of physical distance, because if it too far away, there could be a high latency in the data connection.

There are some pools that offer more than one coin that you can choose from. You can invest in the hardware, connect that up to the pool, and still make quite a nice chunk of change. Or you can just do it on the cheap and get used equipment, especially if you are just getting started. If you can get used PCs, say, from eBay, and then put your own CPUs together and get with the pool, you will find that your investment is low and you can actually increase your rate of return. Buying brand-new GPU cards for between $600 and $1,000 apiece is ok for some people, but you could easily get used desktops for about $50 to $100 each and still make some money.

Chapter 6 Litecoin Mining

Litecoin is about the same age as Bitcoin. LTC came just after BTC and was one of the first altcoins to enter the market. Most of the issues are exactly the same as BTC, including its mining strategy and deceleration of rates. LTC also works on a blockchain, but a separate blockchain from BTC.

The one thing you want to understand as a miner of cryptocurrencies, and the reason why we have chosen to include it in this book, is that while BTC has a hard cap of 21 million coins (that is the most it will ever produce), LTC has a hard cap of 84 million. That means miners will have more to be able to mine, the moment BTC reaches its upper limits.

The good thing also is that the ASICs used for BTC mining can also be used for LTC mining. There aren't many other coins worth mining with ASICs other than these two. The transition is a good way to manage the investment return, going from mining BTC for the next few years and then transitioning to LTC. Or you could just go straight to LCT now, since the miner's universe in BTC seems to be getting really crowded compared to LTC.

At the moment, LTC has 130TH/ day which works out to be 1.5 GH/s vs Bitcoin's 21 EH/s EH is Exahashes. 1 Exa is a billion Giga. So you can see the crowd in Bitcoin compared to the relatively sparse crowd in LTC.

The block time for LTC is approximately 2.5 minutes, meaning that you would need to compute 225 billion hashes to get a block reward. It is still a tall order for an individual with limited hash power, so it is better to join a mining pool and bring in the rewards as a group.

There are a few things you want to consider when you think about mining with LTC. The first is that most people who have the hardware and are mining BTC, are going to move over and come to LTC — once the profitability for BTC starts to go below that of LTC. There are two ways this can happen. One is that the USD/LTC price goes up. It is currently trading at $150 per coin. This is versus $8500 for BTC per coin. Two is that there are no more coins to mine.

Mining LTC is the same as mining BTC, so I don't have to go into the mechanics of it. However, the thing that you have to do is to decide which you will choose, because the investment for both will be the same since you will be using ASIC hardware like the Antminer S9 we talked about in Chapter 1.

Like I said before, you have a decision to make. The reason I've included LTC in this book is that it gives you a good alternative to mine after BTC reaches unprofitable hash rates, or when it reaches closer to 21 million coins. Or, worst-case scenario, the price of BTC comes off its current levels and approaches the value of LTC.

There is only one condition in which you should chose to mine LTC, and that is if you are willing hold the coin that you mine and sell it at a later date when it's higher in value.

If you do it that way, then you get the benefit of a higher mining return in terms of dollars.

Why?

Because the market always corrects itself between the use of hash rates, difficulty and block time — as you learned earlier in the book. The price you get for the coin right now is the entry level, and the market price for LTC has already factored in all these issues. But the one thing that is not factored in is that the miner holds the coins for a later date. The more you hold, the better your return gets over time.

You can also chose to look at your utility payments as a form of annuity that you pay in every month, and then use that as a the investment. Yes, that's right — you have to not only treat your hardware investment as an investment, but also your utility bill that you pay monthly for the running of your rigs. In many cases, miners sell all of their coins as soon as they get them and then pay off their electricity bill with that. But if you hold your coins in anticipation that they will appreciate in value, then you will have to cover the utility bill until the time down the road when you will liquidate your position. In the LTC world, this strategy might make some sense — if you think the price is going to appreciate.

That takes care of the things that you need to know for LTC. The expertise you need to have for BTC will be easily transferable to the mining of LTC, with just the two differences in block times and the cap on the number of coins.

Chapter 7 Zcash Mining

At the time of writing, ZEC is trading at close to $500 per coin. All calculations discussed in this chapter will be based on that figure.

There is one factor of ZEC that interests me more than anything else from among all the other currencies. I am personally a privacy buff. It's not that I intend to participate in anything immoral or illegal, I just value my anonymity and privacy. It doesn't matter what set of eyeballs are looking in at me. It is not just the government's peepers I wish to avoid. And while all the coins claim to be anonymous, they aren't really totally anonymous if you know that to look for. In that way, cash is more anonymous.

You may be wondering what the big deal with anonymity is if you are not doing anything illegal. Let me explain. When you look at the BTC map, you can see where all the coins are and who's got what coin. I may not be able to see a name, but if I sell you something and you pay me from an account, I can see the account balance of that account. That's one problem. It's like if I give you a check, you get to see the total balance in my checking account and all the amounts that have been in there since the time it was created. Yes, there are ways around that, and any BTC holder will tell you to always create a new account when paying someone. Well, the thing is, I can still trace that account and see where that came from. So there is no real privacy. If a hacker wanted to target someone, he just has to go shopping on the map and see where the big balances are.

In ZEC however, they have overcome this issue and made account balances private. Which is a really clever thing to do. ZEC compares to BTC in that they will also have 21 million coins as their hard cap. At present, they have just under 4 million coins in circulation.

ZEC, though similar to BTC and more private, does not need the use of ASICs to mine. You can mine it with GPUs, and as such, realize a healthy return if you join pools. If you follow the investment calculation and use of GPU as we did for Ether, you can get the profitability profile for ZEC as well.

Conclusion

And that brings us to the conclusion of this book. We have looked in detail at the cryptocurrency market and the way to mine them. We've seen the computing issues in mining and the hardware setups that you can use to mine. We have also looked at how to calculate the best course of action, and the various coins that you can focus on with different strategies.

In total, we have looked at how to mine with ASICs, GPUs and CPUs. We've covered whether you should mine alone or in a pool, and noted which coin in particular you should look at.

We even looked to see if you can take a 50% hold of the hash power of the whole market, and drilled down to what the math behind that would look like.

In essence, you now have a wide-ranging understanding of what you can and should do in terms of mining and investing.

There are more than 1200 cryptocurrencies in circulation at various levels of participation. At the top of that list is BTC, and we have looked at that in the most detailed way possible because it is the template for the entire concept of cryptocurrencies. But do not fall in love with BTC, as it is a mature market and this is highly competitive. As such, the profitability is not as aggressive as some of you may like. That being the case, many would argue that the return seems to justify the risk. With that, I would have to say that there is no such thing. Because in any market-driven product, whether it's currencies, landed assets, or equity, all mispricings do eventually return to the median. Don't ever forget that there is always a downside, and past performance is not always a measure of future profitability.

The politics of cryptocurrencies are not discussed in this book because it is directed mainly at those who are thinking of getting into the business of mining and making a living out of it. If you want to get started and can't afford the investment, that is understandable, but you can always start small and work your way up.

But the one thing you must always remember is that this is a business endeavor, no matter how you slice it. You need to look at it as such. You need to see it for what it is and look at the potential as well as the risk — not get all starry-eyed at the potential for all the millions to be made. Or worse — watch YouTube videos of some unknown person living it up because he got into cryptos early in the game. Do not be fooled. This is still a business and it needs to be evaluated with the even-handedness of a professional player.

On a final note, regardless of how you get into mining, the best thing that you can do is start with what you have and work at ploughing the returns back into the business and grow your capacity. There are monetary gains to be made and there are strategic things to consider. We have looked at both in this book.

This book was written in such a way as to give you the exposure to different mining techniques, as well as to help you understand the underlying rationale and logic behind how all of this works. There is still so much that you could dive into, and I suggest you look for books that go deeper into it. If you want to understand cryptography and the other areas that are sprouting up, they will be well worth your time.

Thank You!

Before you go, we would like to thank you for purchasing a copy of our book. Out of the dozens of books you could have picked over ours, you decided to go with this one and for that we are very grateful.

We hope you enjoyed reading it as much as we enjoyed writing it!

We hope you found it very informative.

www.ingramcontent.com/pod-product-compliance
Lightning Source LLC
Chambersburg PA
CBHW081748200326
41597CB00024B/4429